Do It!
Workbook for
NUTRITION
Concepts and Controversies
10 TH EDITION

Frances Sienkiewicz Sizer

Ellie Whitney

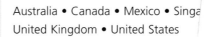

Australia • Canada • Mexico • Singa
United Kingdom • United States

Printed in Canada
4 5 6 7 09 08 07

Printer: Transcontinental Printing/Interglobe

ISBN-13: 978-0-495-01078-4
ISBN-10: 0-495-01078-2

For more information about our products, contact us at:
Thomson Learning Academic Resource Center
1-800-423-0563
For permission to use material from this text or product, submit a request online at
http://www.thomsonrights.com
Any additional questions about permissions can be submitted by email to **thomsonrights@thomson.com**

Thomson Higher Education
10 Davis Drive
Belmont, CA 94002-3098
USA

Asia (including India)
Thomson Learning
5 Shenton Way
#01-01 UIC Building
Singapore 068808

Australia/New Zealand
Thomson Learning Australia
102 Dodds Street
Southbank, Victoria 3006
Australia

Canada
Thomson Nelson
1120 Birchmount Road
Toronto, Ontario M1K 5G4
Canada

UK/Europe/Middle East/Africa
Thomson Learning
High Holborn House
50–51 Bedford Row
London WC1R 4LR
United Kingdom

Latin America
Thomson Learning
Seneca, 53
Colonia Polanco
11560 Mexico
D.F. Mexico

Spain (including Portugal)
Thomson Paraninfo
Calle Magallanes, 25
28015 Madrid, Spain

Contents

 **Analyze Nutrition News**

HERE IS A CHANCE to practice your skills in reading nutrition news with an educated eye. Find an article in your local newspaper or any publication that carries nutrition news. On a copy of Form D1-1, answer these questions:

Step 1. What sort of language does the writer use? Do the words imply sensationalism or conclusive findings? Phrases such as "startling revelation" or "now we know" or "the study proved" are clues to whether the report is a sensational one. Does the author take a tentative approach, using words such as *may, might,* or *could*? What do these words imply?

Step 2. Is the finding placed in the context of previous nutrition science? Does the article imply that the current finding wipes out all that has gone before it? Can you detect a broad understanding of nutrition on the writer's part? From what clues? For example, an article about folate and heart disease should say that saturated fat probably plays the major nutrition role in heart disease development.

Step 3. Does the article mention whether the research results under discussion are published in a medical or nutrition journal? Where? The Controversy section following

Chapter 1 has more information about which types of journals publish valid scientific findings and which do not.

Step 4. How were the results obtained? Can you tell from the article whether this was a case study, an epidemiological study, an intervention study, or a laboratory study? How does that information affect your understanding of what the results have contributed to nutrition science?

Step 5. Does the finding apply to you? Should you change your eating patterns because of it? In what ways did the subjects resemble or differ from you? Were there enough subjects to make the study seem valid? (In a serious evaluation, a statistical analysis would be used to answer this question.)

Step 6. Does the finding make sense to you in light of what you know about nutrition? You may not know enough to make this judgment yet, but by the end of this course, you should have developed a "feel" for identifying information that fits with reality.

This sort of assessment can guide you through the numerous nutrition articles appearing in newspapers and magazines, so you can avoid making nutrition decisions based on passing fads.

FORM D1·1 CRITIQUING NUTRITION NEWS

The news report I am critiquing comes from _____ (publication), dated _____ (attach the report to this form).

1. I evaluate the language used in the publication as follows:

   _____
   _____
   _____

2. I believe the author's understanding of previously reported findings to be:

   _____
   _____
   _____

3. I judge the credibility of the item to be:

   _____
   _____
   _____

4. The methods used to obtain these results were:

   _____
   _____
   _____

5. The results of the study apply to the following populations:

   _____
   _____
   _____

6. To a reader without extensive nutritional background, the results of a study may be misleading. This report might mislead by:

   _____
   _____
   _____

DO IT! Evaluate Nutrient Descriptors on Labels

THIS DO IT! ACTIVITY provides a chance to focus your attention on the meaningful descriptive terms on food labels. Unlike some claims, such as "helps maintain a healthy heart," that may or may not have a firm scientific basis, the nutrient descriptors on food labels, such as "low fat," are well regulated, informative, and reliable. A key reference for this activity is found in Table 2-6 in Chapter 2 of the textbook.

Read the Labels

Part 1 of Figure D2-1 (next page) illustrates some descriptive banners that consumers often encounter on food labels. Part 2 of the figure provides Nutrition Facts panels from food labels that may qualify to bear these descriptors. Read over both parts of Figure D2-1 to get a feel for the nutrients in the foods being described and the potential descriptors. Using the list of descriptive terms in Table 2-6 of the textbook for reference, match up the Nutrition Facts Panels with the correct descriptors on Form D2-1 below. Hint: More than one descriptor may apply to a single Nutrition Facts panel, and more than one Nutrition Facts panel may qualify to bear a particular descriptor. Then answer the questions in the analysis section.

FORM D2·1 MATCHING LABELS WITH DESCRIPTIVE TERMS

Label A _____ _____ _____ _____

Label B _____ _____ _____ _____

Label C _____ _____ _____ _____

Label D _____ _____ _____ _____

Analysis

Answer the following questions:

1. Two of the foods whose labels appear in Figure D2-1 qualify to be called "reduced" in a nutrient. Name the foods and the qualifying nutrients.

2. The requirement to list *trans* fats is relatively new, but it must be taken into account when considering a product's saturated fat content. The American cheese label shown in the figure contains no *trans* fat. Would the FDA therefore allow the words "*trans* fat free" on its label? Read the definition of this term in the text Table 2-6, and then explain why or why not.

3. Just one of the foods qualifies as an "excellent source" of calcium. State which one and explain why it qualifies. Which descriptive term can be applied to the food with the next highest calcium content?

4. Two of these foods provide 5 grams or more of fiber, the qualifying amount to be called a "high fiber" food. Just one of these foods can legally carry the descriptive term on its label, however. Which one? Why did the other food fail to qualify?

5. If you were looking for a "cholesterol free" food among those listed, indicate which two foods would meet your need and explain why the others would not.

Continue practicing this kind of label reading in grocery stores until reading labels becomes a natural part of your food selection process. If you know which parts of the label to take seriously and which to ignore, labels can be a real help in achieving your goal of feeding yourself well.

PART 1

1   2   3   4

PART 2

Tomato Soup

This product contains 450mg sodium vs. 710mg in our regular soup.

Nutrition Facts

Serv. Size 1/2 Cup (120ml) condensed soup
Servings: About 2.5
Calories: 90
Fat Calories: 15

Amount / serving	% DV	Amount / serving	% DV
Total Fat 1.5g	2%	**Potassium** 250mg	7%
Sat. Fat 0.5g	3%	**Total Carb.** 18g	6%
Trans Fat 0g		Fiber 1g	4%
Polyunsat. Fat 0.5g		Sugars 10g	
Monounsat. Fat 0g		**Protein** 1g	
Cholest. 0mg	0%		
Sodium 450mg	19%		

Vitamin A 8% • Vitamin C 10% • Calcium 0% • Iron 2%

A

American Cheese

Saturated fat reduced from 5g to 2g in a serving. Calcium increased from 10% to 25%.

Nutrition Facts

Serv. Size: 1 Slice (21g) Servings: 16
Calories: 50 Fat Calories: 30

Amount / serving	% DV
Total Fat 3g	
Sat. Fat 2g	5%
Trans Fat 0g	10%
Cholesterol 10mg	3%
Sodium 290mg	12%
Total Carb 1g	0%
Fiber 0g	0%
Sugars 1g	
Protein 4g	

Vitamin A 8% • Vitamin C 0%
Calcium 25% • Iron 0%

C

Crunchy Sweet Oatbran Cereal

Nutrition Facts

Serving size	3/4 Cup (49g/1.8 oz)
Servings per Container	About 10

Amount per serving	Cereal	Cereal with 1/2 Cup Vitamins A&D Fat Free Milk
Calories	200	240
Calories from Fat	70	70
	% Daily Value	
Total Fat 8g	13%	13%
Saturated Fat 2g	10%	10%
Trans Fat 1g		
Cholesterol 0mg	0%	0%
Sodium 140mg	6%	9%
Potassium 220mg	6%	12%
Total Carbohydrate 35g	12%	14%
Dietary Fiber 5g	23%	23%
Sugars 15g		
Other Carbohydrate 15g		
Protein 4g		
Vitamin A	15%	20%
Vitamin C	25%	25%
Calcium	2%	15%
Iron	10%	10%

B

Whole Grain Bread

Nutrition Facts

Serving Size: 1 Slice (45g)
Servings: About 16
Calories: 120
Fat Calories: 5

Amount per serving	% DV
Total Fat 0.5g	1%
Saturated Fat 0g	0%
Trans Fat 0g	
Cholesterol 0mg	0%
Sodium 140mg	6%
Total Carbohydrate 26g	9%
Fiber 5g	23%
Sugars 6g	
Protein 4g	

Vitamin A 0% • Vitamin C 0% • Calcium 15% • Iron 10%

D

DO IT! Find the Fiber in Two Lunches

HERE'S A CHANCE to estimate the fiber in two fictional meals so that you can recognize the fiber in your own real-world food choices. You'll need to use three parts of the textbook: Figure 4-15 of Chapter 4, Appendix A at the back of the textbook, and the DRI table on the inside front cover of the book.

Finding the Fiber

Step 1. Read the lunch foods listed in Table D4-1. Take accurate note of the portion sizes: the difference in fiber between a half-cup and a quarter-cup of beans, berries, or any other foods can be sizable.

Step 2. List the fiber grams for each of the foods in both lunches of Table D4-1. Fiber values for many common foods are listed in Figure 4-15 of the textbook; look there first for convenience. For fiber in foods not listed, such as cookies, look in Appendix A of the textbook.

Step 3. Add up the fiber grams to obtain the total for each lunch. Find your DRI recommended intake for fiber (inside text front cover, page C). Now determine what percentage of your day's fiber need each lunch provides. Example:

$$\text{DRI recommendation} = 25 \text{ g}$$
$$\text{Fiber in lunch} = 2 \text{ g}$$
$$2 \text{ g} \div 25 = .08$$
$$.08 \times 100 = 8\%$$

Analysis

Answer the following questions:

1. What percentage of your DRI recommended intake for fiber would be met by each of these lunches?
2. Most authorities suggest obtaining about a third of a day's need for fiber at lunch. By this reckoning, is either of these lunches too high in fiber? Is either too low? Is either just right? Why do you think so?
3. The higher-fiber lunch was constructed of whole foods. List the foods that stand out as rich fiber sources.
4. Read over the foods listed in Figure 4-15 of the text noting the amount of fiber listed for various juices. Why do you think that juices are so much lower in fiber than their whole-food counterparts? What type of fiber do you think is predominant in juice? (Hint: The main constituent of juice is water.)
5. The high-fiber lunch of Table D4-1 provides foods from each of the food groups, but it may include more food than many people care to consume at lunchtime. How can the lunch be reduced in size without greatly affecting the

fiber total and without eliminating any of the food groups? How does the resulting lunch compare with your DRI recommendation for fiber (give the percentage)?

6. Look again at Figure 4-15 of the text. List some foods that you might be willing to include in your diet to help meet your fiber need. To which meals could you add these fiber-containing foods?

TABLE D4·1 A FIBER-RICH LUNCH AND A FIBER-POOR LUNCH

Lunch #1 A Fiber-Rich Lunch

Foods	Fiber (g)
Turkey sandwich:	
• 1 oz. turkey	——g
• 1 oz. Swiss cheese	——g
• 1 tomato slice, ½ inch thick (about ¼ tomato), and a lettuce leaf (about ¼ c)	——g
• 2 slices whole-wheat bread	——g
1 medium pear	——g
Beans in tomato sauce with pork, canned (about ½ c)	——g
8 oz fat-free milk	——g
2 chocolate sandwich cookies	——g
	Total Fiber
	———

$$\frac{}{\text{Total Fiber}} \div \frac{}{\text{DRI}} \times 100 = \underline{\quad} \%$$

Lunch #2 A Fiber-Poor Lunch

Foods	Fiber (g)
Turkey sandwich:	
• 1 oz turkey	——g
• 1 oz Swiss cheese	——g
• 1 tomato slice, ½ inch thick (about ¼ tomato), and a lettuce leaf (about ¼ c)	——g
• 2 slices white bread	——g
1 bag chips (1 oz)	——g
8 oz milk	——g
2 chocolate sandwich cookies	——g
	Total Fiber
	———

$$\frac{}{\text{Total Fiber}} \div \frac{}{\text{DRI}} \times 100 = \underline{\quad} \%$$

Summary

Constructed mostly of refined foods and meats, the diets of many people lack whole plant-derived foods, and the average U.S. consumer fails to consume enough fiber to support health. Does this mean that low-fiber foods such as refined white rolls, pies and cakes, white rice, meats, and apple juice are off-limits? Current recommendations state that half of the day's grain food choices should be whole grains while the remainder may be refined, enriched grains. Further, abundant daily fruits, vegetables, and legumes as part of a balanced diet contribute greatly to fiber intakes. Constructed with these goals in mind, a day's meals can easily meet the recommendation for dietary fiber and include some favorite refined grain foods as well.

DO IT!

Evaluate Saturated and *Trans* Fats on Food Labels

THIS EXERCISE ASKS you to compare lipids in one serving of each of three packaged lasagnas and evaluate them according to intake guidelines. The lasagna labels are found in Figure D5-1 on the next page. One set of guidelines, the Daily Values, is printed right on the label and makes it possible to compare food labels to each other in a meaningful way in the grocery store aisles. Another set, developed by the World Health Organization (WHO), are specific with regard to saturated fat, *trans* fat, and cholesterol and are intended to help healthy people stay healthy (see Table D5-1). The DRI committee urges that both saturated and *trans* fats be kept as low as possible while ensuring adequate nutrient intakes.

Comparing the lipids in foods with intake guidelines requires that you do three things:

- Calculate a personal set of guidelines for saturated fat, and *trans* fat. (Form D5-1, page 9).
- Compare the saturated fat, *trans* fat, and cholesterol among the lasagnas. (Form D5-2, page 9).
- Determine the percentage of each guideline that is delivered by one serving of lasagna. (Section 2, Form D5-2, page 9).

The next sections guide you through these processes and then ask some questions.

In calculating your guidelines for saturated fat or *trans* fat, you will use three numbers. First is your DRI for energy (from the inside front cover of the text). Second is the recommendation to consume no more than 10 percent of calories from saturated fat or 1 percent from *trans* fat. The third number arises from the calorie value of fats: 9 calories for every 1 gram of fat.

Find Your Personal Guidelines

Step 1. Calculate your personal guideline for saturated fat. On Form D5-1, section 1, part A, write in your DRI energy recommendation (inside front cover). Transfer your energy recommendation to part B and calculate 10 percent (the WHO guideline for saturated fat) as shown.

Transfer this number from part B to part C and divide by 9 calories per gram to find grams of saturated fat. The answer is your personal guideline for saturated fat. Copy it into part D for later use.

Step 2. Calculate your personal guideline for trans *fat.* Repeat the process for *trans* fat, using section 2 of Form D5-1.

It's wise to memorize your guidelines for these fats and make food choices that do not exceed them on most days. These are some of the most valuable numbers you can learn.

Compare the Fats in Three Lasagnas

Step 3. Using Form D5-2, enter the grams of saturated fat per serving of lasagna from the three package labels of Figure D5-1. Do the same for the *trans* fat. Then enter your personal guidelines for saturated fat and *trans* fat from Form D5-1. Calculate the percentage of saturated fat and *trans* fat provided by each lasagna as shown.

Analysis

Now use the information you have generated to respond to the following questions:

1. Which lasagna is highest in saturated fat per serving? Which is highest in *trans* fat? What percentage of your personal guidelines for saturated fat and *trans* fat would these lasagnas contribute to your day's total intake?
2. If you ate lasagna high in saturated or *trans* fat, what foods could you choose at other meals to avoid exceeding your recommended maximum intake for the day?
3. If you substituted the lasagna lowest in saturated fat for the one that ranks highest, what effects would this have on your other food choices and on your calorie and nutrient intakes that day?

Now, read the ingredients list for each lasagna and answer the remaining questions:

4. Which ingredients contributed most to the saturated fat in the highest-fat lasagna?
5. Name the ingredients contributing to *trans* fat in each lasagna that contains it.
6. In the lowest-fat lasagna, which ingredients were used in place of high-fat counterparts included in the other lasagnas?

TABLE D5·1 LIPID GUIDELINES FOR CHRONIC DISEASE PREVENTION (WORLD HEALTH ORGANIZATION AVERAGE POPULATION VALUES)

Dietary Factor	Goal (% of total energy)
Saturated fatty acids	<10%
Trans fatty acids	<1%
Cholesterol	<300 mg per day

A

Nutrition Facts

Serving size 10¹/₂ oz (298g)
Servings per Package 1

Amount per serving

Calories 472	Calories from Fat 252

	% Daily Value*
Total Fat 28g	43%
Saturated Fat 16g	80%
Trans Fat 2g	
Cholesterol 125mg	42%
Sodium 820mg	34%
Total Carbohydrate 35g	12%
Dietary fiber 4g	16%
Sugars 9g	
Protein 20g	

Vitamin A 25%	•	Vitamin C <2%
Calcium 50%	•	Iron 10%

*Percent Daily Values are based on a 2,000 calorie diet. Your daily values may be higher or lower depending on your calorie needs.

Daily Values

		Calories	2,000	2,500
Total Fat	Less than		65g	80g
Sat Fat	Less than		20g	25g
Cholesterol	Less than		300mg	300mg
Sodium	Less than		2,400mg	2,400mg
Total Carbohydrate			300g	375g
Dietary Fiber			25g	30g

Calories per gram
Fat 9 • Carbohydrate 4 • Protein 4

INGREDIENTS, Whole Milk, Ricotta Cheese (Whole Milk, Cream, Vinegar and Salt), Cooked Macaroni, Spinach, Cream, Parmesan Cheese, Carrots, Onions, Modified Cornstarch, Bread Crumbs, Corn Syrup, Long Grain Rice Meal, Potato Flakes, Malt, Yeast, Margarine, Vegetable Shortening (Partially Hydrogenated Soybean Oil), Salt, Calcium Propionate, Fat-Free Dry Milk Solids, Salt, Romano Cheese, Mushrooms, Sugar, Salt, Mono- and Diglycerides, Xanthan Gum, Spices, Garlic Salt.

B

Nutrition Facts

Serving size 10¹/₂ oz (298g)
Servings per Package 1

Amount per serving

Calories 361	Calories from Fat 117

	% Daily Value*
Total Fat 13g	20%
Saturated Fat 8g	40%
Trans Fat 0g	
Cholesterol 87mg	29%
Sodium 860mg	36%
Total Carbohydrate 37g	12%
Dietary fiber 0g	
Sugars 8g	
Protein 26g	

Vitamin A 15%	•	Vitamin C 10%
Calcium 25 %	•	Iron 10%

*Percent Daily Values are based on a 2,000 calorie diet. Your daily values may be higher or lower depending on your calorie needs.

Daily Values

		Calories	2,000	2,500
Total Fat	Less than		65g	80g
Sat Fat	Less than		20g	25g
Cholesterol	Less than		300mg	300mg
Sodium	Less than		2,400mg	2,400mg
Total Carbohydrate			300g	375g
Dietary Fiber			25g	30g

Calories per gram
Fat 9 • Carbohydrate 4 • Protein 4

INGREDIENTS, Tomatoes, Cooked Macaroni Product, Dry Curd Cottage Cheese, Beef, Low-Moisture Part-Skim Mozzarella Cheese, Dehydrated Onions, Modified Cornstarch, Salt, Parmesan Cheese, Enriched Wheat Flour, Sugar, Spices, Tomato Flavor (Salt, Tomato Paste and Flavorings), Dehydrated Garlic.

C

Nutrition Facts

Serving size 11 oz (312g)
Servings per Package 1

Amount per serving

Calories 244	Calories from Fat 36

	% Daily Value*
Total Fat 4g	4%
Saturated Fat 2g	10%
Trans Fat 1g	
Cholesterol 10mg	3%
Sodium 500mg	21%
Total Carbohydrate 34g	11%
Dietary fiber 5g	20%
Sugars 10g	
Protein 18g	

Vitamin A 25%	•	Vitamin C 25%
Calcium 40 %	•	Iron 10%

*Percent daily values are based on a 2,000 calorie diet. Your daily values may be higher or lower depending on your calorie needs.

Daily Values

		Calories	2,000	2,500
Total Fat	Less than		65g	80g
Sat Fat	Less than		20g	25g
Cholesterol	Less than		300mg	300mg
Sodium	Less than		2,400mg	2,400mg
Total Carbohydrate			300g	375g
Dietary Fiber			25g	30g

Calories per gram
Fat 9 • Carbohydrate 4 • Protein 4

INGREDIENTS, Tomato Puree, Cooked Enriched Macaroni Product, Soy Protein Isolate, Soy Protein, Guar Gum, Part Skim Ricotta Cheese, Tomatoes, Zucchini, Vegetable Shortening (Partially Hydrogenated Soybean Oil), Carrots, Spinach, Onions, Water, Mushrooms, Concentrated Dealcoholized Burgundy Wine, Sugar, Modified Food Starch, Salt, Spices, Microcrystalline Cellulose, Methylcellulose, Maltodextrin, Hydrolyzed Corn Protein, Xanthum Gum, Guar Gum, Autolyzed Yeast, Calcium Chloride, Citric Acid, Garlic Extractives, Dextrin.

Section 1. Saturated Fat

A. Copy your DRI energy recommendation from page B, inside front cover: _____ calories.
(energy recommendation)

B. Calculate the number of calories you can consume as saturated fat each day:

_____ cal × .1 = _____
(energy recommendation)   (saturated fat calories)

C. How many grams is this?

_____ cal ÷ 9 cal per gram = _____ grams saturated fat
(saturated fat calories from B)

D. Your personal **saturated fat guideline** = _____ g
(from C)

Section 2. *Trans* Fat

A. Copy your DRI energy recommendation from page B, inside front cover: _____ calorires.
(energy recommendation)

B. Calculate the number of calories you can consume as *trans* fat each day:

_____ cal × .01 = _____
(energy recommendation)   (*trans* fat calories)

C. How many grams is this?

_____ cal ÷ 9 cal per gram = _____ grams *trans* fat
(*trans* fat calories from B)

D. Your personal ***trans* fat guideline** = _____ g
(from C)

Section 1. What percentage of your guideline for saturated fat does a serving of each lasagna provide? First, consult the labels in Figure D5-1 and enter the grams of saturated fat in the appropriate spaces below. Also enter your personal guideline for saturated fat in the appropriate lines and then calculate the percentages as indicated.

Lasagna A _____ g ÷ _____ g × 100 = _____ % of personal guideline for saturated fat
(saturated fat)   **(saturated fat guideline)**

Lasagna B _____ g ÷ _____ g × 100 = _____ % of personal guideline for saturated fat
(saturated fat)   **(saturated fat guideline)**

Lasagna C _____ g ÷ _____ g × 100 = _____ % of personal guideline for saturated fat
(saturated fat)   **(saturated fat guideline)**

Section 2. What percentage of your guideline for *trans* fat does a serving of each lasagna present? Repeat the process described above, using *trans* fat values.

Lasagna A _____ g ÷ _____ g × 100 = _____ % of personal guideline for *trans* fat
(*trans* fat)   **(*trans* fat guideline)**

Lasagna B _____ g ÷ _____ g × 100 = _____ % of personal guideline for *trans* fat
(*trans* fat)   **(*trans* fat guideline)**

Lasagna C _____ g ÷ _____ g × 100 = _____ % of personal guideline for *trans* fat
(*trans* fat)   **(*trans* fat guideline)**

CONSIDER THE SOURCES of protein in a day's meals. Look at the meals in Figure D6-1. Breakfast and lunch are given, but supper is yet to be planned. A simple breakfast of cereal, milk, and juice provides 14 grams of protein. Lunch is a bit heartier with a ham and cheese sandwich contributing most of its 18 protein grams. Now comes a puzzle—which supper to choose? After picking a supper from among the choices, check Figure D6-2 on the next page to find out how much protein each supper provides (the totals are printed upside down to prevent you from peeking before you guess). Then compare the protein in the meals with your DRI protein intake recommendation (use the "grams per day" values on the inside front cover of the text, page A).

Two of the supper options are meatless (the spaghetti supper and the vegetable-rice supper), and one contains meat. To quickly assess the protein in such meals, remember that fruits provide negligible protein, but meats, milk, and cheeses are the richest sources, followed by legumes, grains, breads, and vegetables.

Analysis

Settle on a supper choice and consider these questions:

1. Did the supper you chose contain enough protein so that the day's total meets, but does not exceed, your protein recommendation? Which other suppers also qualify?
2. If your answer to the first part of question 1 was no, which foods might you substitute to achieve your goal without shorting yourself on nutrients? Hint: If your protein exceeded your goal, you may need to restrict

---

FIGURE D6·1   A PROTEIN PUZZLE*

The breakfast and lunch of this day's meals have contributed 32 grams of protein to the eater. What would you choose for supper? Supper A features spaghetti and tomato sauce; supper B is based on sautéed vegetables and brown rice; supper C offers two small lamb chops (2 oz each). Figure D6-2 on the next page reveals their protein contents.

© Polara Studios, Inc. (all)

Breakfast	Protein Grams
Orange juice, 1 c	2 g
Cheerios cereal, 1 oz	4 g
Low-fat mlk, 1 c	8 g
Total	14 g

Lunch	Protein Grams
Iced tea,	0 g
Sandwich	
2 slices whole wheat bread,	
1 slice lunch ham,	
1 slice cheese,	
¾ c lettuce and tomato	17 g
Peaches, ½ c	1 g
Total	18 g

Supper A = ? protein

Supper B = ? protein

Supper C = ? protein

*The protein values listed in this exercise are from the *Food Processor Plus, 7.11,* a computerized diet analysis software program developed by ESHA Research.

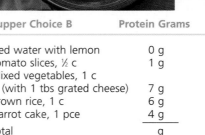
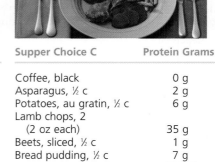

© Polara Studios, Inc. (all)

Supper Choice A	Protein Grams
Iced water with lemon	0 g
Garlic bread, 2 pieces	6 g
Salad (with ¼ c each garbanzo beans, artichoke, and cucumber	6 g
Spaghetti, 1 c (with 1 tbs parmesan cheese)	13 g
Sherbet, 1 c	2 g
Total	g

Supper A total = 27 g protein
Entire day's protein = 59 g

Supper Choice B	Protein Grams
Iced water with lemon	0 g
Tomato slices, ½ c	1 g
Mixed vegetables, 1 c (with 1 tbs grated cheese)	7 g
Brown rice, 1 c	6 g
Carrot cake, 1 pce	4 g
Total	g

Supper B total = 18 g protein
Entire day's protein = 50 g

Supper Choice C	Protein Grams
Coffee, black	0 g
Asparagus, ½ c	2 g
Potatoes, au gratin, ½ c	6 g
Lamb chops, 2 (2 oz each)	35 g
Beets, sliced, ½ c	1 g
Bread pudding, ½ c	7 g
Total	g

Supper C total = 51 g protein
Entire day's protein = 83 g

something, and an obvious "something" to restrict is some of the meat.

3. Make some educated guesses concerning the other two energy-yielding nutrients, fat and carbohydrate. Which foods contribute abundant fat and saturated fat to this day's meals? Which contribute carbohydrates and fiber?

4. Note that breakfast, though it contains no meat, provides almost as much protein as the ham and cheese sandwich at lunch. Which foods in this breakfast provide protein? Describe how the amino acids in some foods might complement those of other foods in the breakfast.

5. Which plant food shown in Figure D6-2 is the richest in protein? Which is next richest? Hint: If you have eliminated or are considering eliminating meat from your diet,

read Chapter 6's Controversy section—it points out the pros and cons of both vegetarian and meat-containing diets.

6. If you were to design a day's meals around the lamb supper, yet did not want to consume too much protein, how would you change breakfast and lunch? What foods would you substitute for some of the protein-rich foods listed for the breakfast and lunch in Figure D6-1? Another hint: If you are considering doing away with the milk or cheese, remember that you must then provide other sources of calcium (you may reconsider this decision when you discover in Chapter 8 that few foods other than milk products or specially calcium-fortified foods supply an abundance of calcium).

DO IT! Find the Vitamins on a Menu

CAN YOU MEET your vitamin needs when you eat in restaurants? You can if you learn to identify the foods on restaurant menus that are rich sources of vitamins. Assume you are spending a day on the road and have to eat all three of the day's meals in restaurants. Read over the three menus in Figure D7-1 and create a meal from each one. Just like real menus, these menus lack serving size information. For this exercise, take for granted that the serving sizes of foods agree with the standard-size servings listed in Figure 2-4 in Chapter 2 of the textbook. (Never assume this about real menus, however—commercial serving sizes vary enormously and can be several times larger than standard servings.) In Figure D7-1, most vegetables and grains are in ½-cup servings. Those identified as "large" are 1-cup servings; meats are 2- to 3-ounce portions; milk is an 8-ounce serving.

Step 1. On a copy of Form D7-1 (next page), record your food choices down the left-hand column.

Step 2. Consult Table 7-5 of the textbook to determine the values for calories and nutrients in the day's meals. Fill in the calorie and nutrient values for the foods you listed on Form D7-1. Coffee, tea, and water contribute negligible energy and vitamins; use zeros for their values on Form D7-1.

Step 3. Enter the DRI intake recommendations that apply to your age and gender (see the inside front cover of the textbook) in the spaces provided at the bottom of Form D7-1.

Step 4. Divide the day's total intakes by the recommendation amounts and multiply by 100 to determine the percentages of your nutrient and energy needs contributed by the foods chosen this day.

- Example: If total vitamin C intake equals 40 milligrams and your vitamin C recommendation equals 90 milligrams, then $(40 \div 90) \times 100 = 44\%$. The meals provided less than half of the daily recommended amount of vitamin C.

Repeat this process for all five vitamins and for energy.

FIGURE D7·1 RESTAURANT MENUS

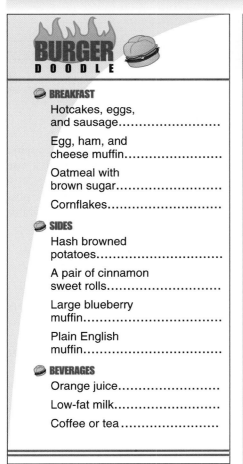

BURGER DOODLE

BREAKFAST
- Hotcakes, eggs, and sausage......................
- Egg, ham, and cheese muffin......................
- Oatmeal with brown sugar......................
- Cornflakes...........................

SIDES
- Hash browned potatoes.......................
- A pair of cinnamon sweet rolls..........................
- Large blueberry muffin................................
- Plain English muffin...............................

BEVERAGES
- Orange juice........................
- Low-fat milk........................
- Coffee or tea......................

The Box Lunch Express

Soups
Homemade chili with crackers ....

Sandwiches
Cold cut hoagie sandwich with chips ....

Peanut butter and jelly sandwich on whole-wheat bread, served with fruit cocktail ....

Tuna sandwich on white bread, served with a fresh banana ....

Salads
Large chef's salad with cheese, ham, and turkey with crackers ....

Beverages
Fat-free milk ....
Apple juice.....................................
Sparkling water.............................

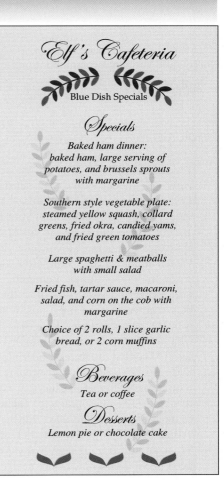

Elf's Cafeteria

Blue Dish Specials

Specials

Baked ham dinner:
baked ham, large serving of potatoes, and brussels sprouts with margarine

Southern style vegetable plate:
steamed yellow squash, collard greens, fried okra, candied yams, and fried green tomatoes

Large spaghetti & meatballs with small salad

Fried fish, tartar sauce, macaroni, salad, and corn on the cob with margarine

Choice of 2 rolls, 1 slice garlic bread, or 2 corn muffins

Beverages
Tea or coffee

Desserts
Lemon pie or chocolate cake

	Energy (cal)	Vitamin A (μg)	Thiamin (mg)	Vitamin C (mg)	Folate (μg)	Vitamin B₁₂ (μg)
Breakfast:						
Lunch:						
Supper:						
Day's totals						
Energy and vitamin recommendations						
% of recommendations						

Analysis

Answer the following questions:

1. How did the day's totals for calories compare with your DRI energy recommendation? Did the day's meals provide too much, or too little, of your need for energy? If the answer is too much or too little, what changes in your choices would correct the problem?

2. Which foods listed on the menus contribute significant amounts of vitamin A? Remembering that brightly colored fruits and vegetables offer the vitamin A precursor beta-carotene, name any foods in your meal choices that provide beta-carotene (turn back to Snapshot 7-1 in Chapter 7 of the text for hints). Which foods that you selected contribute little or no vitamin A in any form?

3. Did your choices supply enough folate to meet your requirement? Remembering that young adults are urged to obtain 400 micrograms of folate each day, what percentage of your folate need did the day's meals provide? Which individual foods were richest in folate?

4. What about vitamin C? What percentage of your daily need of vitamin C did your meals provide? Which foods were the main contributors?

5. When considering vitamin B₁₂, which food groups stand out as good sources? Which individual foods that you selected were the main contributors?

6. Which foods on the menus, though low in vitamins, possess other valuable constituents that make them desirable as part of a health-promoting diet?

If you are wondering about the vitamins in other foods you choose, you can find their vitamin and other nutrient contents in several references in the textbook:

- The vitamin and mineral Snapshots appearing throughout Chapters 7 and 8 depict the richest vitamin and mineral sources.
- Figure 2-4 in Chapter 2 notes which vitamins characterize foods of each food group.
- The Table of Food Composition, Appendix A at the back of the textbook, provides actual values for nutrients in thousands of foods.

With these references, you can judge the vitamin values of foods on virtually any menu.

 **DO IT!** Find the Minerals in Snack Foods

IN AN IDEAL WORLD, you would set aside time every day for planning, shopping, and cooking nutritious meals. In the real world, you've had nothing to eat, your research papers are due, your room is topsy-turvy, and your car needs fuel as you're rushing to class. A convenience store seems like a good idea; you can fill up your car and grab a bite to eat in one stop. What you grab makes a difference to your day's calorie and mineral intakes, however—to the benefit or detriment of your nutritional health.

The labels on munchies list their contents of energy, calcium, iron, and sodium. We've done the same in Table D8-1 on the next page.

List Your Nutrient Intake Goals and Choose Some Foods

Step 1. On a piece of paper, list your intake goals for energy, calcium, and iron (see the DRI, inside front cover of the textbook). For sodium, use the recommended upper limit of 2,300 milligrams as your target for the day.

Step 2. Scan the variety of items on the convenience store shelves represented in Figure D8-1. Choose a snack and check off your choices in the spaces provided. The portion sizes are those commonly used for such foods, and not those recommended in the Food Guide Pyramid. For example, a can of Vienna sausages contains 5 ounces of sausages, not the recommended 3 ounces.

Record Energy and Mineral Values

Step 3. Obtain energy and mineral values for your snack choices from Table D8-1. On your piece of paper, total the values for energy and minerals in your snack.

Step 4. Calculate the percentages of energy and mineral requirements contributed by this snack. Divide the totals for energy and each nutrient by the recommendation for that nutrient, and multiply by 100. Example: For a snack providing 84 milligrams of calcium with calcium intake recommendation of 1,000 milligrams:

$$(84 \div 1,000) \times 100 = 8.4\%$$

The snack would meet more than 8 percent of your DRI recommended intake for calcium.

Analysis

Now answer the following questions:

1. What percentage of your energy need did the snack contribute? Did the snack also contribute a proportional amount of needed minerals? Which ones?
2. What about sodium? Did the sodium in this snack exceed 10 percent of the maximum of 2,300 milligrams? Experts

---

FIGURE D8·1 SNACKS AT THE FUEL 'n' FEED

Snacks at the *Fuel 'n' Feed*

✓ **Sandwiches/ Canned foods**
—Cheese pizza slice
—Ham biscuit
—Pork and beans
—Roast beef sandwich
—Sardines
—Vienna sausage

✓ **Beverages**
—Milk, 1% fat
—Chocolate milk, 2% fat
—Cola
—Pineapple orange juice

✓ **Snacks**
—Apple pie, packaged
—Banana
—Cheese slice
—Chocolate candy bar
—Dill pickle
—Dried fruit mix
—Fig bar cookies
—Ice cream and sherbet bar
—Potato chips, small bag
—Pretzels, regular
—Pretzels, unsalted
—Roasted almonds
—Sunflower seeds, shelled
—Wheat crackers
—Yogurt, low fat, with fruit

---

suggest that a serving of food should supply less than 10 percent of the sodium for the day. Did the snack exceed 30 percent? If so, suggest some foods to choose at other meals to prevent exceeding the sodium maximum for the day. Consult Figure 8-8 of the textbook to review the principles of high- and low-sodium foods.

3. Which of the foods listed in Table D8-1 are the best "bargains" in terms of providing less sodium and more calcium and iron?
4. Identify any iron "bargains" you find. List snack foods rich in needed iron, yet relatively low in calories.
5. Can a person seeking calcium do well in a convenience store? Name the snack foods on your list that supply substantial calcium (10 percent of the recommendation) and compare their calorie values. If your choices lack calcium, find and list other snacks that provide it.

Quick snacks are actually small meals that contribute to a day's intakes of energy and nutrients. Chosen with nutrition in mind and consumed in the context of a nutritious diet, many snacks can make valuable contributions to the day's nutrient needs.

TABLE D8·1   ENERGY AND MINERAL CONTENTS OF SELECTED CONVENIENCE STORE FOODS

√ Check your snack choices	Energy (cal)	Calcium (mg)	Iron (mg)	Sodium (mg)
_____ Cheese pizza, slice	243	184	0.7	467
_____ Ham biscuit	386	160	2.7	1,432
_____ Pork and beans, 5 oz	139	80	4.7	624
_____ Roast beef sandwich	318	47	3.1	1,252
_____ Sardines, 3¾ oz	221	405	3.1	536
_____ Vienna sausages, 5 oz	395	14	1.3	1,347
_____ Milk, 1% fat, 1 c	102	300	0.1	123
_____ Low-fat chocolate milk, 2% fat, 1 c	179	285	0.6	151
_____ Cola, 12 oz	186	14	0.1	18
_____ Pineapple orange drink, 12 oz	170	17	0.9	10
_____ Apple pie, fast-food type	225	6	1.0	179
_____ Banana	105	7	0.4	1
_____ Cheese slice, 1 oz	69	121	0.2	250
_____ Chocolate candy bar, 1½ oz	226	84	0.6	36
_____ Dill pickle	12	6	0.3	833
_____ Dried fruit mix, 3¾ oz	258	40	2.9	19
_____ Fig bar cookie, 4	195	36	1.3	214
_____ Ice cream and sherbet bar	92	62	0.1	43
_____ Potato chips, salt and vinegar, 1 oz	150	7	0.5	380
_____ Pretzels, salted, 1 oz	108	10	1.2	486
_____ Pretzels, unsalted, 1 oz	110	10	1.2	60
_____ Roasted almonds, salted, 3¾ oz	660	326	4.0	828
_____ Sunflower seeds, shelled, 3¾ oz	654	123	7.2	3
_____ Wheat crackers, 1 oz	134	14	1.3	225
_____ Low-fat yogurt, with fruit, 8 oz	232	345	0.2	133

DO IT! — Control the Calories in a Day's Meals

THIS EXERCISE PROVIDES practice in controlling calorie intakes. Figure 9-14 of the textbook demonstrated how to reduce calories in a day's meals to save over 1100 calories a day. Using the foods in Figure 9-14 of Chapter 9, try your hand at cutting calories further to 1,800 or even 1,600 calories for the day.

The only "must" in cutting calories is to make the diet adequate. Be sure to include required amounts of food from each of the food groups to maintain nutrient adequacy. Make substitutions with care: substituting diet cola for one of the two cups of milk would cut calories but would also compromise calcium adequacy and so is not allowable. Remember to make at least half the grain choices whole grains and to include several teaspoons of unsaturated oil to provide vitamin E and essential fatty acids.

Step 1. Use Form D9-1, the column marked *Lower-Calorie Choices,* to record your changes in the lower-calorie day's meals of text Figure 9-14 (already listed for convenience at the left of the form).

Step 2. Record the calorie savings. Turn to the Table of Food Composition, Appendix A of the textbook, to find calorie values for foods you propose as substitutions for the

FORM D9·1 TRY YOUR HAND: REDUCE CALORIES FURTHER

2,300-Calorie Day	Lower-Calorie Choices	Calories Saved	Food Group Amount	
			Name of Food Group	Amount (cups or ounces)
Breakfast				
Fat free milk, 1 c, 83 cal				
Orange juice, 1 c, 112 cal				
Whole grain waffle, 1 each, 201 cal				
Soft margarine, 1 tsp, 34 cal				
Syrup, 2 tbs, 105 cal				
Banana slices, ½ c, 69 cal				
Lunch				
Fat free milk, 1 c, 83 cal				
Cheeseburger, small, 2 oz, 330 cal				
Green salad, 1 c,				
with 1 tbs light dressing, 67 cal				
croutons, ½ c, 50 cal				
French fries, regular (about 30), 210 cal				
Ketchup, 1 tbs, 16 cal				
Gelatin dessert sugar free, 20 cal				
Dinner				
Italian bread, 1 slice, 81 cal				
Soft margarine, 1 tsp, 34 cal				
Stewed skinless chicken breast, 4 oz., 202 cal				
Tomato sauce, ½ c, 40 cal				
Rice, 1 c, 216 cal				
Mixed vegetables, ½ c, 59 cal				
Low-fat cheese sauce, ¼ c, 85 cal				
Brownie, 1 each, 224 cal				
Totals 2,321 cal		Calories saved:		

Amount from Each Food Group:
Milk, yogurt, cheese ﹍﹍ Fruits ﹍﹍ Vegetables ﹍﹍
Meats and legumes ﹍﹍ Grains ﹍﹍ Oils (tsp) ﹍﹍

2,321 − ﹍﹍﹍﹍﹍ = ﹍﹍﹍﹍﹍
        (calories   (new day's
        saved)   total calories)

originals. There is no need to look up every food on the menu—just find substitutes for those you choose to change. Subtract the calorie value of the new food from that of the original and write the calorie difference in the *Calories Saved* column.

Step 3. Add the *Calories Saved* column to obtain your total calorie savings for the day.

Step 4. Subtract the total savings from the original total of 2,321 calories to find the calorie value of the day's new meals.

Step 5. Assign each food from the new menu to its appropriate food group. Write the food group names on the left-hand side of the *Food Group* column of Form D9-1. In the column to the right, titled *Amount,* list the amount of each food in cups or ounces (equivalent values for foods, for example ounces of cheese to cups of milk, are listed in Figure 2-4 in Chapter 2 of the textbook). Total the day's amounts from each group, and write the totals in the spaces provided at the bottom of the form.

Note that while oils are not an official food group, they are necessary in the amounts suggested in Table 2-3 of Chapter 2 of the textbook. Total up the teaspoons of oil in soft margarine, vegetable oil, salad dressings and other sources. Hint: the oil in fried foods does not qualify in this accounting because vitamin E is destroyed at high temperatures.

Amounts

Answer the following questions:

1. How many total calories did your changes save?
2. Assuming that a pound of body weight is worth 3,500 calories, how much weight could a dieter theoretically lose in a month by cutting every day's calories to this extent?
3. Did you remove high-calorie constituents from any foods? Which ones?
4. Which high-calorie foods did you replace with lower-calorie ones? Try to judge how these changes may have affected the saturated fat, vitamin, mineral, and fiber content of the meals; write your responses.
5. Were you able to cut calories significantly while still meeting the minimum number of servings from each food group? If not, which groups fell short? List ways of adjusting your choices to include the missing foods.
6. Did your meals include any sweets or other treats? If so, which ones? If not, why not?

When you develop skill in making these sorts of changes, they tend to come to mind whenever the opportunity arises. Choosing foods with an eye for their contributions of calories and nutrients becomes a natural part of living.